contents

what is evolution?

A ccording to the latest estimates, there may be as many as 30 million species – different kinds of living things – on Earth. Life permeates our planet, from the lower reaches of the stratosphere, about 12km (7 miles) high, down to the deepest oceans. But how did life start in the first place? And how did all these different living things come to be as they are today? For the vast majority of scientists, there is only one satisfactory explanation: they evolved. Evolution is a gradual process of chemical and physical change that began even before fully-fledged life appeared, and that still continues now. It has left its imprint in everything that is or was once alive, including our distant ancestors and ourselves. It is responsible for the way we look, the way we reproduce, and – some would argue – even the way we behave. Evolution is biology's "big idea". Ushered into life nearly 150 years ago, it is more relevant than ever in today's changing world.

" Nothing in biology makes sense except in the light of evolution. "

T. Dobzhansky, geneticist, 1973

evolution

DAVID BURNIE

LONDON, NEW YORK, MUNICH,
MELBOURNE, DELHI

**DORLING KINDERSLEY,
LONDON**

senior editors Peter Frances and Hazel Richardson
DTP designer Rajen Shah
picture researcher Rose Horridge
illustrator Halli Verrinder

category publisher Jonathan Metcalf
managing art editor Philip Ormerod

Produced for Dorling Kindersley Limited by
Cobalt id, The Stables, Wood Farm, Deopham Road,
Attleborough, Norfolk NR17 1AJ, UK

First published in Grea
Dorling Kinder
80 Strand, Londo

A CIP catalogue record for th
the British

ISBN 0-7513-7355-9

Colour reproduction by Colourscan, Singapore
Printed and bound by
Graphicom, Italy

See our complete catalogue at
www.dk.com

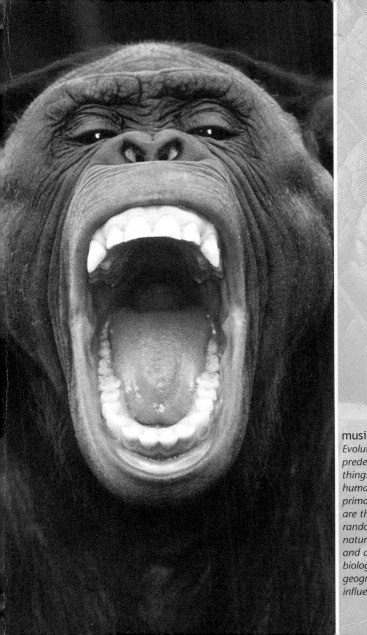

music of chance
Evolution is not predestined. All living things, including humans and our primate relatives, are the products of random mutation, natural selection, and a range of other biological and geographical influences.

evidence from the past

early ancestor
This skeleton belonged to Procyonosuchus, *a terrier-sized animal that lived 250 million years ago.* Procyonosuchus *was a member of a group of reptile-like animals called therapsids, which gave rise to the earliest mammals.*

During the 1790s, Thomas Jefferson – the future president of the United States – unearthed a set of animal bones in Virginia. Despite their odd shape and large size, he decided that they must have belonged to a lion, rather than to an animal that no longer existed. In Jefferson's day, extinction was a controversial subject, because accepting it seemed to imply that divine creation was somehow flawed. Today's biologists take the opposite view: extinction is a natural feature of life. Fossilized remains, like those of Jefferson's "lion" (actually a giant sloth) have been excavated and catalogued in their millions. They show that far from staying the same, living things have always undergone change, and that huge numbers of them have died out.

windows on the past

Fossils are laid down and preserved in layers of rock, and so form a chronology of life on Earth. Unfortunately, this record is riddled with gaps. That is because the chances of something being successfully fossilized, then discovered millions of years later, are extremely slim. And the fossil record is biased because some things, such as animals with shells or hard skeletons, fossilize much more easily than others. But despite these shortcomings, it contains a vast amount of information about life in the past. Even tiny fragments of fossils can throw light on species that are long extinct. In 1999, a piece of

trace fossils

Most fossils are the preserved remains of organisms, but some are indirect remains, created by the activities of living things. Among them are fossilized droppings, bite marks, footprints, trackways (collections of footprints), and burrows. These so-called trace fossils give rare clues about how extinct animals behaved. Many trace fossils are microscopic, but some trackways cover hundreds of square kilometres. The spacing and arrangement of individual prints can show how fast an animal moved, and whether or not it lived in groups.

hominid footprint
This fossilized footprint is from a set of tracks left by a human ancestor called Australopithecus afarensis. *It is about 3.6 million years old, and shows that this hominid walked upright.*

jawbone – together with three minute teeth – was unearthed in Madagascar. Microscopic examination revealed that the teeth had a complex shape, with sharp points and cutting edges – characteristic of insect-eating mammals. The animal they belonged to, *Ambondro mahabo*, turned out to be more than 165 million years old, making it one of the most ancient mammals known. Key fossil finds like *Ambondro* help to show how existing groups of living things evolved, and provide an epitaph for groups that have died out. They can even throw light on the conditions and climate in the distant past.

ancient evidence
Stromatolites are rocky mounds made by bacteria that live in shallow seas. Their fossils date back over 3 billion years, making them one of the first forms of life on Earth.

fossilization

When something becomes fossilized, its remains partly escape decay. Soft tissues break down, but hard parts of the body, such as bones and shells, become mineralized, preserving their original shape. They can stay buried for millions of years until erosion brings them back to the surface. Whether or not the remains become fossilized is a matter of chance; the process is easily disrupted. This diagram shows a successful fossilization sequence, with examples of what can "go wrong" at each stage to the right.

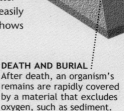

DISINTEGRATION
The organism dies but is not covered up. Its remains are eaten by scavengers, or destroyed by weathering.

DEATH AND BURIAL
After death, an organism's remains are rapidly covered by a material that excludes oxygen, such as sediment.

MINERALIZATION
Any soft tissues slowly decompose. Meanwhile, hard parts, such as bones or bark, are gradually infiltrated by waterborne minerals.

COMPRESSION
The mineralized remains become buried more and more deeply under layers of accumulating sediment. Time and pressure turn the mineralized material into rock.

trapped in amber

This spider became trapped in resin oozing from a tree several million years ago. The resin later became fossilized (amber), preserving the spider within. Remains can also be preserved by mummification (drying out), and by burial in permanently frozen ground.

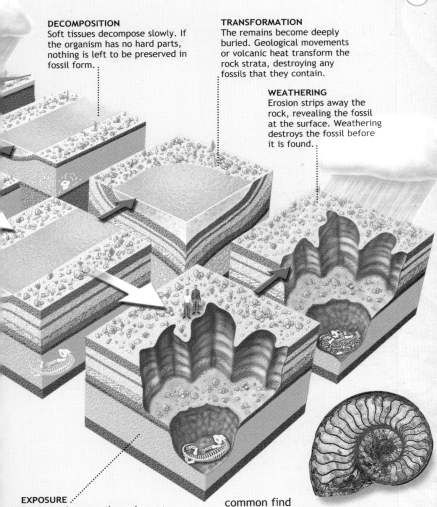

DECOMPOSITION
Soft tissues decompose slowly. If the organism has no hard parts, nothing is left to be preserved in fossil form.

TRANSFORMATION
The remains become deeply buried. Geological movements or volcanic heat transform the rock strata, destroying any fossils that they contain.

WEATHERING
Erosion strips away the rock, revealing the fossil at the surface. Weathering destroys the fossil before it is found.

EXPOSURE
Erosion strips away the rock, revealing the fossil at the surface. A keen-eyed paleontologist finds the fossil before it is broken up by the processes of weathering.

common find
Ammonites were widespread in the world's oceans between 400 and 65 million years ago. They are amongst the most commonly found fossils.

living evidence

If fossils did not exist, it would be extremely difficult to trace the path followed by evolution. Even so, there would still be plenty of evidence to confirm that evolution had actually taken place. That evidence is built into the fabric of living things, and consists of underlying similarities and unusual "quirks" of design or behaviour that are extremely unlikely to have come about by chance. Compared to fossils, this kind of evidence is easy to get hold of, but interpreting it is not always straightforward.

significant feet
Frigate birds have part-webbed feet, usually associated with water birds. However, they collect their food on the wing, and hardly ever land on water. Their feet are an evolutionary legacy, inherited from ancestors that behaved more like typical seabirds.

shared patterns

Unlike a human designer, evolution never starts with an entirely blank drawing board. Instead, it works by modifying things that already exist. As time goes by, new changes build upon older ones, but long-established patterns are very slow to disappear. If two living things share one of these patterns, there is a strong likelihood that they are related, and that they have inherited it from the same source. These shared patterns, or homologies, are powerful evidence for evolutionary change. One striking example – studied by Charles Darwin – involves the pattern of bones in mammalian limbs (see right). Throughout the living world, homologies crop up again and again, linking species that often have quite different ways of life.

Without evolution, the natural world would be static, and living things would always have followed the lifestyles that they have now. But many living species carry the signs of earlier lifestyles, in the form of body parts that no longer have any real use. These "vestigial organs" include a host of redundant structures, such as the eyes of cave-dwelling fish, and the rudimentary hind limb bones of some snakes and whales. In our own case, they include the appendix, which once played a part in digestion. Evolution also explains some anatomical oddities that seem like simple bad design. The giraffe's nervous system is one. Like all vertebrates, it has some

homologies

Homologous structures occur in organisms that share a common ancestor. Over time, they evolve in quite different directions, but a shared pattern remains. For example, dolphins and chimpanzees evolved from a common ancestor: their limbs work in very different ways, but they have the same underlying arrangement of bones. Analogous structures are ones that carry out the same function, but that evolved from different starting points – wings of insects and birds, for example.

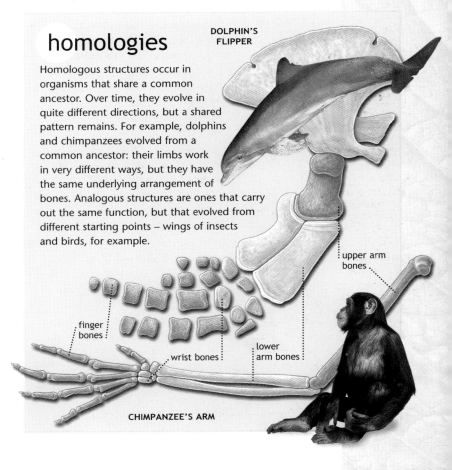

DOLPHIN'S FLIPPER

upper arm bones

finger bones

wrist bones

lower arm bones

CHIMPANZEE'S ARM

> ❝ What can be more curious than that the hand of a man, formed for grasping, that of a mole for digging, and the wing of a bird should all be constructed on the same pattern... ❞

Charles Darwin, 1859

nerves that run from its brain down to its chest, and then back up to its larynx. This there-and-back-again route was inherited from fish, which have no necks at all. In fish, it made sense, but in the giraffe it involves several metres of extra wiring.

The evolutionary leftovers carried by living organisms include physical features, internal processes, and aspects of behaviour. For example, until the age of about three months, human infants show a powerful gripping reflex, and can even support their own weight with their hands. This almost certainly dates back to the time when our new-born ancestors clung to their mothers by their fur.

living links

The living world contains some species that seem to bridge the gap between major groups of organisms. For example, a handful of mammals – the duck-billed platypus and echidnas, found only in Australia and New Guinea – lay eggs, just like their reptilian ancestors. They are the most primitive mammals alive today. Similarly, some fish resemble amphibians in breathing air through their lungs. These living "halfway houses" provide strong hints about the path that evolution has followed in the past.

platypus
The platypus is a very odd mammal. It has webbed feet and uses its duck-like bill to grub in the mud for its prey of soft-bodied invertebrates.

what makes things evolve?

Taken together, the evidence from fossils and from homologies in living things makes it clear that change does indeed take place. However, identifying change is one thing; deciding what drives it is quite another. In the early 1800s, the French naturalist Jean-Baptiste de Lamarck suggested that change was driven by living things themselves, as they strove to perfect their way of life. For example, in Lamarck's words, a wading bird "makes its best efforts to stretch and lengthen its legs", and as a result, increasingly long legs become a feature of its descendants.

Superficially, it sounds plausible. But as Lamarck's critics pointed out, living things are not capable of this kind of planning, and even if they were, the idea has another significant flaw: features acquired during an individual's lifetime are not passed on to its offspring. If they were, trees growing in windy places would produce stunted saplings, and parents who do weight-training would have muscle-bound children.

in competition

During the 1830s, the young English naturalist Charles Darwin completed a round-the-world-voyage aboard the naval survey ship HMS *Beagle*. By the end of the five-year trip, Darwin had amassed a wealth of evidence for evolution, although he did not yet know why it took

alive
A shield bug guards her newly-hatched young. Like all living things, these insects have the potential to produce far more offspring than their environment can support, creating competition for limited resources.

Charles Darwin (1809–1882) did not "discover" evolution, nor was he the only person to come up with the idea of natural selection. His achievement was to marshal the evidence for both in a conclusive and comprehensive way. Most of the observations that informed his theories were made during a five-year voyage as naturalist aboard the HMS *Beagle*.

place. Darwin's great breakthrough came in 1838, when he read an essay on the growth of the human population. Its writer, the English economist and cleric, Thomas Malthus, argued that humans have a natural tendency to outstrip their food supply, creating competition for scarce resources. Darwin immediately grasped the huge significance of this idea: competition constantly takes place in nature as well, giving rise to a permanent struggle for survival.

winners and losers

From the observations he made on his travels, Darwin knew that living things show a host of inherited variations. He realized that in any struggle for resources, some variations – or characteristics – must prove more useful than others. The owners of these "winning" features would leave larger numbers of offspring, and as a result, their characteristics would gradually become more widespread in the population as a whole. The end

❝ It is not the strongest species that survive, nor the most intelligent, but the ones most responsive to change. ❞

Charles Darwin, 1859

result is change, driven by a passive process that he called natural selection. Unlike Lamarck's version of evolution, Darwin's involves no planning or preset goals. In any species – from bacteria to elephants – individuals are "judged" by one simple criterion: their ability to leave the most young that survive to reproduce.

genes and variation

When he wrote *On the Origin of Species*, Darwin had no idea how features were passed from one generation to the next. Plant breeding studies carried out by his contemporary, Gregor Mendel, showed that characteristics are carried by discrete "factors" which are derived from one parent or the other. In 1909, these basic units of hereditary material were named "genes", and 20th century scientists devoted much energy to elucidating their physical nature. Today, we know that genes are sequences of four chemical bases, (abbreviated to C, G, A, and T) that are "written into" the length of molecules of deoxyribonucleic acid (DNA), contained within the chromosomes of every cell. When an organism reproduces, DNA copies itself, and the parental genes are passed on. The copying process is accurate, but mistakes, or mutations, do sometimes occur.

DNA
The DNA molecule, which is shaped like a double helix, contains a vast library of coded commands that direct an organism's growth and development.

mutations
Some mutations produce characteristics that improve an organism's chances of survival, but most have negative effects, such as misforming the eye of this fruit fly (right).

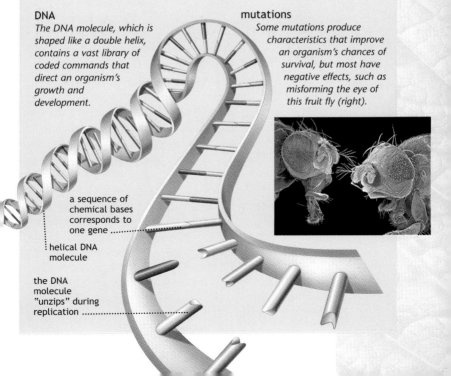

a sequence of chemical bases corresponds to one gene ....

helical DNA molecule

the DNA molecule "unzips" during replication ....

hand picked

The natural variability of plants and animals can be exploited to serve our own ends through artificial selection. Choosing to breed from parents with desirable characteristics can create new forms of crop plants and domestic animals. The wild cabbage is a classic example: selective breeding has given rise to a number of very different crops.

kale
selection for leaves

brussels sprouts
selection for lateral buds

broccoli
selection for stems and flowers

kohlrabi
selection for stems

cabbage
selection for terminal buds

cauliflower
selection for flower clusters

ancestral
Brassica oleracea

plastic shapes

Genes have ultimate control over the size and shape of a plant or animal, but these features can also be modified by environment. A leaf grown in shade is broader to catch more light.

oak leaf grown in shade

leaf grown in light

Darwin was a meticulous worker, and he spent the following two decades preparing his extensive research for eventual publication. But in 1858, he discovered that he was about to be scooped. Another English naturalist – Alfred Russel Wallace – had also hit on the idea of natural selection, although he had much less research to back it up. This promoted Darwin to begin writing in earnest. The result, one year later, was *On the Origin of Species* – the best selling science book of all time.

how natural selection works

Natural selection is a powerful yet extremely subtle process. It is constantly at work, testing all the minute variations among living things, and favouring any that enable them to leave more young. Unlike deliberate, or artificial, selection, it does not have any preferred outcomes or ultimate aims, and it cannot encourage features that might be useful in the future unless they are already advantageous now. That means that every current feature of living things has evolved through an almost infinite series of steps, each of which helped its owners to survive.

what is "fit?"

Natural selection is often summed up as "survival of the fittest" – a phrase first coined by Herbert Spencer, a 19th-century philosopher. It is a neat summary of Darwin's key idea, but in evolutionary terms, it does not have quite the same meaning as it does in everyday life.

singled out
Predators like this shark are important agents of natural selection. They tend to catch the least fit individuals among their prey, leaving the fittest individuals to breed. As a result, the fittest individuals leave the most young.

sexual selection

A special form of selection occurs in species where individuals of one sex – usually the males – compete with each other to attract mates. In this contest, any adaptation that successfully attracts females is an advantage, because it allows a male to father the most young. This process, known as sexual selection, explains why some male animals have eyecatching "accessories", such as flamboyant plumage or extra-large antlers. Sexual selection can favour these adaptations even if they are a handicap in daily life. Over many generations, it tends to make males and females more and more different – a situation called sexual dimorphism.

headgear
The extinct Irish elk had gigantic antlers that spanned nearly 4m (13ft).

That is because the "fittest" individuals are not necessarily the strongest or fastest. In some cases, fitness can come through what look like negative qualities, such as being small and timid rather than big and bold.

moving goalposts

Another feature of Darwinian fitness is that it does not necessarily stay the same. Something that is fit today may not be quite so fit tomorrow, because the environment has somehow changed. Natural selection does not linger over past favourites: if they are at a disadvantage, they are sidelined without delay. Persistent losers leave fewer and fewer offspring, so their features become more and more rare. Eventually, they become extinct – natural selection's final thumbs-down.

It is easy enough to imagine how natural selection can favour changes that produce immediate benefits, such as bigger muscles or warmer fur. But even Darwin had difficulty imagining how it could create complex organs, such as human eyes. Nevertheless, he was convinced that eyes evolved in a series of small steps – a view shared by biologists today.

Eyes certainly look as if they have been created intact, because all their parts work together as a whole. But a close look at the animal kingdom shows that there is a wide spectrum of eye "design": some eyes are simple, others are complex, but crucially, they all work. By using any one as a springboard, natural selection could make one that is still more sophisticated – a powerful argument for evolution step-by-step.

caution: reversing

Natural selection emphasizes features that help in the survival stakes, but it also minimizes ones that are no longer pulling their weight. A striking example involves the wings of birds. For birds, flying involves a major investment in muscle power – and it is one that normally pays off in the struggle to survive. But if a bird evolves that feeds and nests on the ground, and has few significant predators, it no longer needs to fly. In these conditions, natural selection favours smaller flight muscles and wings, eventually creating species that can no longer fly. This process is the same one that produces vestigial organs – ones that no longer have a useful function at all.

dead weight
Adaptations that were once useful can become a burden. In birds, when flight ceases to be useful, it becomes a serious handicap.

surviving fire
In dry habitats, fire is a powerful agent of natural selection, because it quickly eradicates anything that cannot tolerate being burned. This eucalyptus forest looks doomed, but its trees are fire-adapted. They will sprout fresh leaves a few weeks after the fire has passed.

effects of selection

In the real world, natural selection can have three different outcomes. The most common kind, called stabilizing selection, favours individuals that possess "average" charactistics, and penalizes ones that lie at the extremes of what is normal. This kind of selection reduces the amount of variation in a species, making it more homogenous as time goes by. The second type is called directional selection. This favours individuals

STABILIZING SELECTION

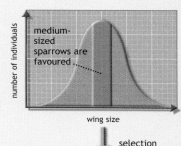

medium-sized sparrows are favoured

number of individuals

wing size

selection

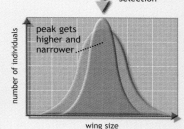

peak gets higher and narrower

number of individuals

wing size

at one extreme.
Over the generations, it makes the average shift towards this extreme, until an optimum point is reached and stabilizing selection then sets in. The third and rarest form of selection favours opposite extremes simultaneously, penalizing individuals that lie in between. Known as disruptive selection, it can split a species into genetically distinct groups. If it continues for long enough, it leads to the creation of new species.

In sparrows, wing size and body weight are both maintained by stabilizing selection. Individuals with medium-sized wings are the most energy-efficient, so natural selection favours them. This kind of selection is typical of long-established species.

DIRECTIONAL SELECTION

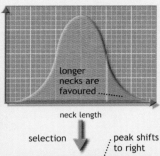

longer necks are favoured ....

neck length

selection ⬇

peak shifts to right

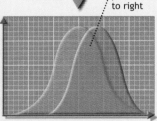

neck length

Giraffes with long necks avoid competing for food with shorter-necked individuals. Directional selection acts to increase average neck length. When its positive effects start to be outweighed by anatomical disadvantages, stabilizing selection takes over.

DISRUPTIVE SELECTION

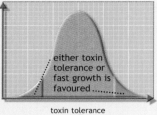

either toxin tolerance or fast growth is favoured ....

toxin tolerance

selection ⬇

two peaks form

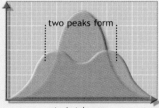

toxin tolerance

Disruptive selection has created two different types of grasses. One type has a tolerance to heavy metal toxins in the soil and grows only on old mine workings. The other grows more rapidly, but has no toxin tolerance. Intermediate forms do well in neither habitat.

evolution in action

In the natural world, many processes are fast enough to be tested in real time. For example, it is a simple matter to confirm the pull of gravity – all you have to do is drop something and watch it hit the ground. But evolution is different. Years may go by between one generation and the next, so the effects of evolution may take hundreds of years to become evident. As a result, evolution is difficult to verify by normal scientific methods, making it that much easier for sceptics to dismiss. But while evolution itself is rarely seen, its outcome is all around. It shows up in the huge range of adaptations found in living things, from their outward shape to the complex ways in which they interact. As these adaptations build up, new species come into existence. At the same time, ones that fail in the struggle for survival eventually become extinct. In the history of life, this cradle-to-grave sequence has been repeated millions of times. Its past record is stored in fossils – a data bank that shows some interesting variations in the speed of evolutionary change.

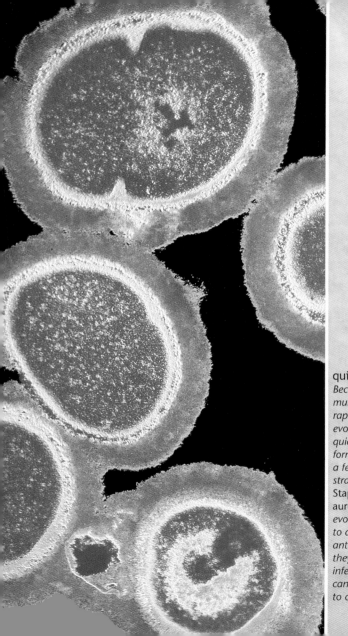

quick change
Because bacteria multiply very rapidly, they can evolve far more quickly than higher forms of life. In just a few decades, strains of Staphylococcus aureus have evolved resistance to almost all antibiotics, and they now cause infections that can be very difficult to control.

adaptation

trigger happy
The Venus flytrap has evolved sophisticated traps that catch unwary insects. Each one is a highly modified leaf, with two sides joined by a hinge. If an insect touches hairs on the leaf surface, the trap snaps shut, and the victim is then slowly digested.

on the move
Migration is an adaptation that allows animals to exploit more than one kind of habitat. In the salmon family, some species migrate every year, but others migrate just once and die after breeding.

An adaptation is any kind of inherited change that improves an organism's chances of survival. The most striking adaptations are anatomical, and it is no accident that these initially helped the theory of evolution to become established. But in evolution, change can affect any feature controlled by genes, from shape and size to instinctive behaviour. Given enough time, almost everything is up for grabs, including adaptability itself. Adaptations look like purposeful changes, and this is sometimes how they are portrayed. For example, it is often said that horses evolved long legs so they could run faster, or that polar bears evolved white fur so they could hide against the snow. Statements like these are a handy shorthand, but they are misleading: adaptations happen as a result of

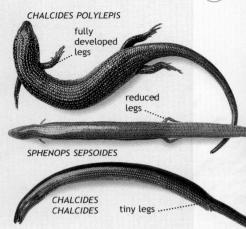

shrinking skink

Skinks are small lizards that feed in loose soil, where wriggling is a more effective way of moving than running. Over generations, some species have evolved reduced limbs, and many have lost them altogether. Limb reduction or loss has evolved repeatedly in skinks and other lizards in many different parts of the world – a strong indication that this adaptation brings advantages in the struggle for survival.

CHALCIDES POLYLEPIS
fully developed legs

reduced legs

SPHENOPS SEPSOIDES

CHALCIDES CHALCIDES   tiny legs

natural selection, not because living things "decide" that they would be useful.

For all living things, maintaining shape is a number one priority, and a key area of adaptation. For larger land-based organisms staying in shape means having some kind of support. Over millions of years, this has led to some very different physical adaptations in animals, including hard outer shells and body cases, and internal skeletons made of cartilage and bone. Plants have solved the same problem in a different way: their structural support comes mainly from overlapping layers of cellulose that are wrapped around every cell, and this adaptation can support trees weighing thousands of tonnes.

changing track

Physical adaptations that originally emerged for one purpose often take on other functions during the course of evolution. For example, the earliest vertebrates (animals with backbones) were primitive fish that did not have jaws. Jaws developed later, evolving from bony struts

preadaptation

Sometimes, through sheer good luck, a species can find that it already has features that enable it to make the most of a change in its environment. These features are called preadaptations. A common example is tolerance of climatic extremes. If the climate turns colder or drier, a preadapted species will soon become more widespread. Weeds are plants that are preadapted to a landscape altered by humans. Because their natural habitat is disturbed ground, they can rapidly spread to and colonize ploughed farmland and other marginal areas.

new homes for old
Swallows originally nested in rock crevices and tree hollows. They were perfectly preadapted to exploit structures built by humans. Today, most swallows nest in barns, roofs, or under bridges.

that originally supported the animal's gills. Before jaws evolved, fish could only eat things that were small enough to suck up whole. But equipped with jaws, they became able to bite off pieces of food, opening up many new ways of life. Leaves are another case of changing roles. Originally, their sole function was to harvest energy from sunlight, but over time, some plants evolved leaves specialized for different purposes. For example, many of the traps used by carnivorous plants are leaves that have become adapted for catching prey.

cacti
The spines of a cactus are actually modified leaves. They give the plant protection against herbivores, and their small size cuts down on water loss in arid habitats. The job of photosynthesis has been taken over by the cactus's fleshy stem.

coping with extremes

Many adaptations involve the internal processes that keep an organism alive. These adaptations are particularly apparent in extreme environments, where living things need extraordinary abilities to survive. In regions with cold winters, many animals hibernate – an adaptation that allows them to go without food at a time when it is difficult to find. At the opposite end of the spectrum, some desert animals have evolved an extremely high tolerance of drought. But nature's true "extremophiles" show some truly mind-boggling

capabilities. Microscopic water animals called water bears, or tardigrades, can survive almost total dehydration for over a decade, while some bacteria can live in hot springs that are more acidic than vinegar. Their adaptations are the result of intense "selection pressure" – an extra-rigorous weeding-out that takes place in tough habitats.

closing down
Hibernation is a metabolic adaptation that cuts an animal's energy use by nine-tenths.

behavioural adaptations

For many biologists, behaviour is one of the most intriguing aspects of adaptation – partly because it has implications for ourselves. At its most basic, behaviour involves simple reflexes – the kind shown, for example, when a snail senses danger and shrinks into its shell. But it also includes some extremely complex activities, such as courtship, nest-building, and migration. Despite their sophistication, most kinds of animal behaviour are instinctive, and because instinct is coded by genes, it can evolve.

pathfinders
Navigation in birds is amongst the most astonishing of all behavioural adaptations.

Like other adaptations, behaviour sometimes betrays signs of its own evolutionary past. For example, when birds court their partners, they often ruffle their feathers with their beaks. This piece of behaviour has almost certainly evolved from the preening procedure they use to keep their feathers clean. Many male birds also present the females with food – a ritual that has evolved from behaviour that is normally used when feeding young. Male insects also do this, although some offer an empty silk cocoon, rather than food itself. While the female unwraps her gift, the male takes the chance to mate.

a neutral stance

Adaptations are everywhere in the living world. So does this mean that every feature of living things has some "adaptive value", no matter how well hidden, or how small? Many evolutionary biologists think that the answer is a definite yes, but a significant number disagree. The

convergence

When unrelated species share the same way of life and face the same challenges to survival, natural selection often throws up similar solutions. Not surprisingly, the species often come to look more alike – a phenomenon known as convergence. A classic example is found among ant-eating mammals, which share numerous feeding adaptations – long snouts, tongues, and powerful claws with which to dig up ants' nests. The phenomenon of convergence highlights the fact that there are a limited number of "engineering" solutions to the problems posed by particular habitats and lifestyles.

anteater
South American tropics

shared shapes
The five mammals shown here all feed partly or exclusively on ants or termites. They are only very distant relatives, and widely spread geographically, but they share similar adaptations that help them to deal with their tiny prey.

dissenters suggest that many genetic mutations have no effect – positive or negative – on an organism's fitness, so that natural selection neither favours them nor acts against them, as long as conditions do not change. If this is true, the characteristics that they create cannot be true adaptations, because they do not help in the struggle for survival. Some features may be no more than by-products of an animal's design, with no adaptive value of their own. A good analogy from architecture is the spandrel – the "V" shaped area that inevitably exists between two adjacent arches. The arches perform a real function – holding up the roof; the spandrel between them has no role, but it is there nevertheless.

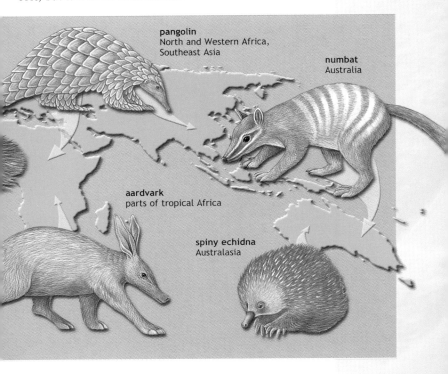

pangolin
North and Western Africa,
Southeast Asia

numbat
Australia

aardvark
parts of tropical Africa

spiny echidna
Australasia

evolving together

With the exception of some microorganisms, none of Earth's natural inhabitants live entirely on their own. Instead, the vast majority are surrounded, or even infested, by other forms of life. Over the long term of evolutionary time, these neighbours influence each other in many different ways. The outcome is coevolution: a process of change in one species that is triggered by changes in another. Coevolution takes place in a wide range of partnerships, some loose, others extremely close.

interactions

When two species coexist in the same habitat, the interactions between them may be positive, negative, or neutral. These relationships are not static, but evolve and change character over time. Positive–negative interactions in which one party gains and the other loses are abundant, particularly in the animal kingdom. They include predation and parasitism. Positive–positive interactions, known as mutualism, are also widespread. They typically

predation
This is a strictly one-way relationship, with the prey (here a gazelle) providing the predator (cheetah) with a source of food.

parasitism
Most parasites behave like predators in miniature. They feed on or within their hosts, but generally do not kill them.

But no matter how close partners may seem, each species in the partnership is ultimately out for itself.

the eaters and the eaten

For most of the world's living things, the need to eat – or to avoid being eaten – is a powerful driving force behind evolutionary change. It may seem odd to think of predators and their prey as "partners", but their influence on each other's evolution can be extremely strong. Their relationship is like an arms race, with one side developing ever more effective ways of catching, or fending off, the other. During this race, natural selection constantly pushes each partner to its limits to generate the best weapons or defences. Interestingly, it is a contest that predators can never totally "win". If they did, their food would vanish, along with their own chances of survival.

coral
Relatives of sea anemones, corals are animals that grow fast in shallow, nutrient-poor tropical waters. They can grow rapidly because their cells contain algae that make sugar from sunlight; the algae then pass some of this food to their animal hosts.

mutualism
A bee gathers pollen and nectar, and enables a plant to reproduce. Pollination is an important case of mutualism.

commensalism
A clownfish benefits by sheltering from predators amongst a sea anemone's stinging tentacles: the anemone gains nothing in return.

involve two very different species – for example, an animal and a plant – that exploit each other in a mutually beneficial way. A relationship in which one participant benefits but the other is unaffected is called commensalism (literally, "eating from the same table"). Negative–negative interactions are common, occuring wherever two animals or plants compete for the same resources. Neutral interactions, where neither partner is affected, are thought to be rare, and are hard to verify.

cuckoo
Dwarfing its foster parent, a young cuckoo swallows a delivery of food. This kind of reproductive trickery, where an animal fools another species into rearing its young, is called brood parasitism. It has evolved in about one per cent of bird species, but many birds occasionally lay eggs in others' nests.

parasites and hosts

In another kind of relationship – parasitism – one partner plunders the other's resources, and often uses its body as a home. Some parasites have just one kind of host, but many have two or more, and live in quite different ways within each host. During their life cycles, parasites often produce enormous amounts of young – a vital adaptation if the chances of reaching a new host are low. Some parasites kill their hosts, but in evolutionary terms, this dead-end situation cannot last for long. Instead, the parasite and its host(s) gradually adapt to one another, and the host species, though weakened, continues to survive.

living together

In a third, and very common, type of partnership, called symbiosis or mutualism, both partners benefit, although not necessarily in equal shares. By exploiting each other's abilities, they fare better than they would on their own. Some kinds of mutualism are "optional", because the partners can still lead independent lives, but over the long term, coevolution often locks the partners together, so they become totally interdependent. This is the case with many flowering plants and their insect pollinators, and also with plant-eating mammals and the microbes within their guts, which help them to digest their food.

Throughout the history of life, alliances like these have proved to be tremendously effective. Without them, about nine-tenths of the world's present species would not exist.

> **❝ Evolution is, for most of the time, a race to stay in the same place. ❞**
> Steve Jones, geneticist, 1999

the ultimate partnership

The cells of animals and plants (eukaryotic cells) contain structures called organelles, which carry out specialized tasks. Mitochondria, for example, are organelles that produce the cell's chemical energy, while chloroplasts harness light energy to make sugar. Biologists have noted that these organelles are remarkably similar to some free-living bacteria, and believe that their early ancestors were almost certainly free-living organisms. About two billion years ago they joined forces with primitive cells to form the first eukaryotic cells.

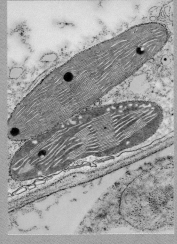

chloroplasts

Plant cells use their chloroplasts to make sugars. With their elliptical shape and layered internal structure, chloroplasts are very similar to cyanobacteria – simple free-living organisms.

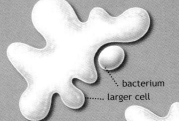

.... bacterium
.... larger cell

ancestral
cell engulfs
bacterium

origins

Eukaryotic cells were formed when ancestral cells engulfed free-living bacteria. Organelles betray this legacy: they have their own DNA, and when their host cells divide, they "reproduce" in parallel.

eukaryotic cell

how new species evolve

During the history of life, perhaps as many as five billion different species have come into being. Most of these are now extinct, but the process of species formation still goes on, and will continue wherever inherited variation exists. Despite this, speciation – as it is known – is a rare event, and the mechanisms that lie behind it are the subject of current research.

isolation
New species often form when a small group of individuals is split off from its parent population by an accident of geography.

what is a species?

If you ask anyone to list a dozen different species of animal, they are unlikely to have trouble understanding what you mean. But in scientific terms, defining the word "species" is surprisingly tricky. At one time, naturalists distinguished species solely on the basis of their morphology – through differences that could be seen. Today, biologists add an important rider: the idea of reproductive isolation. According to this updated definition, known as the "biological species concept", a species is a group of similar organisms that breed only – or almost only – with themselves, and that produce fertile offspring. In genetic terms, this means that they have their own private assortment of genes. This assortment is gradually altered by selection and other forces, sending the members of the species down an evolutionary pathway that is uniquely their own. One interesting aspect of this definition concerns the passage of time. No one can bring together a modern rabbit and one that lived 10,000 years ago, but in theory, they would probably be able to interbreed. This means that they would belong to the same species. But winding the clock back further takes us to a point in time when they could not. At this stage, a threshold has been crossed: species A, the modern rabbit, has developed features that make it distinct from species B, the prehistoric rabbit. This kind of speciation is called transformation, or anagenesis. It changes species but does not increase their numbers.

splitting apart

When Charles Darwin wrote *On the Origin of Species*, he concerned himself mostly with this kind of change. But there is an even more important type of speciation, which occurs when one species gives rise to two or more descendants. Unlike modern and prehistoric rabbits, these new species co-exist in time and can be directly compared. These comparisons usually reveal small outward differences, for example in the shape of leaves, or

periodic insects
Periodical cicadas are North American insects that spend most of their lives below ground. Some emerge to breed once every 13 years, and some every 17 years. These two types are always "out of step" and so cannot interbreed, thus forming separate species.

Darwin's finches
Charles Darwin saw more than a dozen kinds of finches in his visit to the Galápagos Islands. Each had a characteristic beak shape and distinct way of life. He later concluded that the finches had all evolved from a single pioneer species – a process called adaptive radiation.

speciation by geography

Geographical separation is thought to be the most important driving force behind species formation. It occurs when one interbreeding population becomes fragmented into smaller subpopulations. Natural selection changes each one in different ways. If they later meet, their genetic differences may make it difficult or impossible for them to interbreed. By definition, this means that new species have formed.

This kind of separation can be triggered by a variety of geological processes. The slowest but most far-reaching is continental drift, which can break up existing landmasses, splitting up their animal and plant populations. If fragments of continents become remote islands, the stage is set for speciation on a massive scale. This is how Australia and Madagascar came into existence, and it explains why they

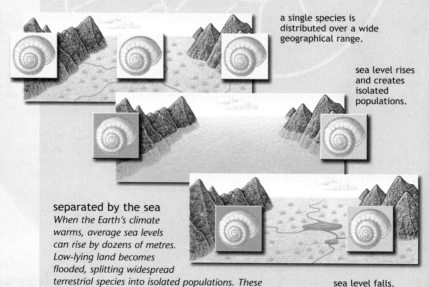

a single species is distributed over a wide geographical range.

sea level rises and creates isolated populations.

separated by the sea
When the Earth's climate warms, average sea levels can rise by dozens of metres. Low-lying land becomes flooded, splitting widespread terrestrial species into isolated populations. These populations follow separate evolutionary paths because each one adapts to its own local environment. By the time sea levels fall again, and the populations can once more come into contact, they may no longer be able to interbreed.

sea level falls. Populations have become distinct species and can no longer interbreed.

have so many endemic species. The reverse process – continental collision – can also split species by building mountains that form barriers between populations.

Climate change is another, more rapid, cause of separation. When the climate warms, polar ice melts and seawater expands, making average sea levels rise. The rising seawater then creates barriers that some terrestrial species cannot bridge. For aquatic species – particularly ones that live in lakes and shallow seas – the opposite is true. Falling water levels split them into isolated groups, each surrounded by dry land. Trapped in these landlocked habitats, they often evolve in very different ways.

AFRICAN LAKES

Lake Victoria

Lake Tanganyika

Lake Nyasa

Aulonacara nyassae

Melanochromis chipokae

Pseudotropheus zebra

Pseudotropheus livingstonii

LAKE NYASA

Melanochromis johanni

Pseudotropheus lombardoi

Rift Valley fish

Water levels in Lake Nyasa in East Africa's Rift Valley have risen and fallen periodically over the last two million years. Land barriers have formed and disappeared, regularly splitting and reuniting populations of fish in the lake. This regular geographic isolation has acted like a "species pump" creating more than 1,500 species of related cichlids.

Scale
0   90 Km
     56 miles

end of the line
Most isolating mechanisms prevent species interbreeding, but some work in a different way, by reducing the viability of hybrid young. Horses and donkeys, for example, can breed successfully, but their offspring – known as mules – are infertile.

female horse

male donkey

mule

the colour of fur. But more interestingly, they often highlight isolating mechanisms – adaptations that make sure species keep themselves to themselves.

These mechanisms occur in many guises. Some involve physical differences or biochemical disparities that prevent two species breeding. In animals, they can also involve differences in behaviour that prevent males from one species attracting females from another. For example, some species of songbirds look so similar that they are very difficult to tell apart. However, the males' songs are completely different – a behavioural adaptation that means that they only attract their own kind.

Differences like these normally evolve when a single species becomes split into separate groups. Once separated, each group or population then sets off down its own evolutionary path. If the groups later come together, some of their adaptations may act as isolating mechanisms, preventing interbreeding. This process is known as geographical, or allopatric, speciation. It explains why remote habitats, such as islands, mountain ranges, and deep lakes, often have a large proportion of endemic species – ones found there and nowhere else.

sympatric speciation

Geographical separation is the most important factor in the creation of new species. But species can also evolve sympatrically – without physical separation. For this to happen, natural selection must create differing populations that overlap but that do not interbreed.

Despite many investigations, such gradual splitting has proved difficult to pinpoint in the wild, but there are plenty of examples of it happening instantaneously in plants. It occurs when genetic "accidents" change chromosome numbers, creating individuals that cannot breed with their parent species.

changing chromosomes

Most mutations occur at the molecular level and change just a single gene. However, sometimes they can affect entire chromosomes – the structures that contain a cell's DNA. One of the most far-reaching of these mutations occurs when a whole set of chromosomes is accidentally duplicated during cell division. This kind of accident, called polyploidy, has played an important part in the evolution of many crop plants.

doubling up

Durum wheat, with a set of 28 chromosomes, probably arose as a polyploid hybrid between two wild grasses, each with 14 chromosomes. Bread wheat evolved when durum wheat hybridized. Each increase in chromosome numbers produced larger plants with greater yields.

chromosome set

All species have a characteristic number of chromosomes, which are usually present as a diploid, or double, set. Chromosome numbers vary widely. Fruit flies, for example, have a total of 8, humans 46, and dogs 78. Some plants have several hundred.

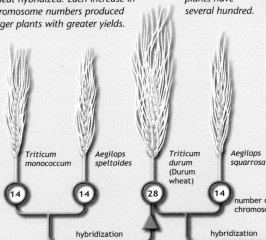

Triticum monococcum — 14

Aegilops speltoides — 14

Triticum durum (Durum wheat) — 28

Aegilops squarrosa — 14

Triticum aestivum (Bread wheat) — 42

number of chromosomes

hybridization

hybridization

time and change

The fossil record does not just show what lived on Earth in the distant past. It also indicates when different species came into existence, and when they disappeared. By studying this fossilized history, scientists can get some idea about the overall tempo of evolution, and whether or not it varies over time. Over the last three decades, this research has prompted a long-running scientific debate.

leap or crawl?
The pattern and tempo of evolutionary change is hotly debated by biologists. Evolution may proceed slowly and gradually, or in sudden leaps of change followed by stasis.

changing rates

In any species, the rate of evolution is controlled by many different factors. They include external ones, such as changing competition from other species, and also an inbuilt one – the rate at which new mutations appear. The inbuilt rate varies from one species to another, but even within a species, it can change from time to time. This is because some genes increase mutation rates, while others slow them down. As a result, the overall speed of evolution can "wobble", just like the price of shares.

micro- and macroevolution

The majority of biologists think that small, gradual changes (microevolution) eventually add up to the major changes that create new species (macroevolution). And because microevolution is gradual, although wobbly, macroevolution should be as well. Many sections of the fossil record appear to bear this out, but others seem to be at odds with this idea. For example, in one key piece of research, 21 species of East African snails were tracked back over a period of more than four million years. For most of this time, they changed remarkably little, but on a few occasions this apparent calm was interrupted by a sudden burst of speciation. On a global scale, far bigger bursts of speciation seem to be a regular feature of life.

In the early 1970s, two American paleontologists – Stephen J. Gould and Niles Eldredge – suggested that this stop-go pattern is not the exception but the rule. According to this idea, known as the "punctuated equilibrium" model of speciation, macroevolution happens in rapid bursts, separated by long periods of

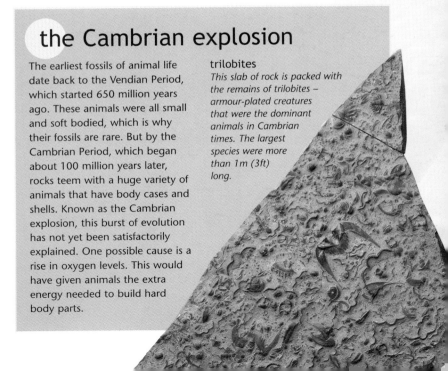

the Cambrian explosion

The earliest fossils of animal life date back to the Vendian Period, which started 650 million years ago. These animals were all small and soft bodied, which is why their fossils are rare. But by the Cambrian Period, which began about 100 million years later, rocks teem with a huge variety of animals that have body cases and shells. Known as the Cambrian explosion, this burst of evolution has not yet been satisfactorily explained. One possible cause is a rise in oxygen levels. This would have given animals the extra energy needed to build hard body parts.

trilobites
This slab of rock is packed with the remains of trilobites – armour-plated creatures that were the dominant animals in Cambrian times. The largest species were more than 1m (3ft) long.

slow change
Horseshoe crabs have changed little in over 400 million years. Despite this, they show as much genetic variability as species that have evolved much more rapidly.

famous fish
Before the discovery of living specimens, 65 million-year-old fossils like this were thought to mark the end of the coelacanth line.

equilibrium. Some "punctuationists" have even proposed that macro- and microevolution are driven by quite different forces. Outwardly at least, some fossil sequences do support the idea of evolution in sudden bursts. But for many paleontologists, the crux of the matter depends on what "sudden" actually means. In the case of the East African snails, the burst of speciation took place over a period of perhaps 50,000 years – more than enough time for gradual evolution to bring new species into being. With global bursts of speciation, the timescale often involves several million years – an even more generous expanse of time.

living fossils

Today, there is widespread agreement that "conventional" microevolution can account for waves of speciation, particularly during times of rapid environmental change. But the flip side of punctuated equilibrium – that evolution can slow to a crawl – has also provoked plenty of study. With species known as living fossils, evolution certainly looks as if it has stopped, but even here, things are not quite what they seem.

For many years, the coelacanth was known only from its fossils, which dated back some 350 million years. Then, in 1938, a living coelacanth was caught off South Africa; and in the late 1990s, the scientific world was astonished by the discovery of a second species of coelacanth off Indonesia. Although the two species look almost identical, genetic analysis has shown differences between them. For these ancient fish, the evolutionary clock may have slowed, but it is still ticking.

extinction

Natural selection brings about some remarkable adaptations, but not even the most impressive can guarantee permanent success in the struggle for survival. Over the long term even the most prosperous species can find themselves in decline, until they eventually become extinct. Many species become extinct on their own, after being undermined by their own limitations or failure to adapt. But sometimes large numbers go together – the victims of catastrophic environmental change.

different ends

In evolutionary terms, extinction can take two different forms. In the first – extinction by transformation – a species evolves so much that it effectively replaces itself. Although the original species is technically extinct, most of its genes live on in the new species that it has produced. This is the kind of extinction that almost certainly befell *Homo erectus* – the species that was our own immediate ancestor. But in the most common form of extinction, something much more final occurs. A species dies out, taking its complete gene pool with it. As a result, its evolutionary line comes to an irreversible end.

dodo
Extinction is increasingly linked to human activity. The dodo, a large flightless bird found only on the islands of Mauritius and Réunion, was wiped out by hungry sailors within 70 years of the islands' discovery in 1598.

impact
The Barringer Crater, near Winslow, Arizona, was created about 50,000 years ago, by a meteorite that weighed at least two million tonnes. An impact of this size would have had devastating effects for life throughout North America.

In the fight for survival, absolute numbers are not that important. This is because some species are viable with a small total population, particularly if they live in isolated habitats, such as oceanic islands. What matters far more is the long-term population trend. Zero population growth keeps a species on course for survival, although it leaves no margin for safety if conditions change. On the other hand, even for an abundant species, a drop in numbers can spell disaster if it is maintained for long enough.

mass extinctions

Extinction normally occurs at a low background rate, as existing species die out and are replaced. But on at least five occasions in the history of life, it has happened on a vast scale. The most famous of these mass extinctions was the one that saw the end of the dinosaurs, but even greater extinction events have occurred at points further back in the past. Each of these mass extinctions was triggered

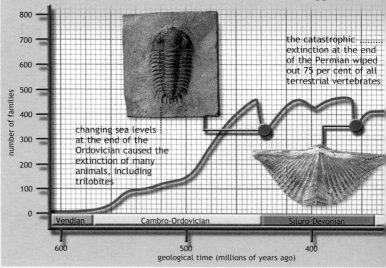

the catastrophic extinction at the end of the Permian wiped out 75 per cent of all terrestrial vertebrates

changing sea levels at the end of the Ordovician caused the extinction of many animals, including trilobites

number of families

geological time (millions of years ago)

Vendian   Cambro-Ordovician   Siluro-Devonian

This danger is illustrated by the fate of the North American passenger pigeon. At one time, this was perhaps the most numerous bird in the world, with a population of 10 billion. The reproductive success of the species depended on it nesting in huge colonies, surrounded by potential mates. So when large-scale hunting began in the 1800s, the colonies began to shrink. Small falls in reproductive rates turned into a headlong decline, and the last survivor, called Martha, died in a zoo in 1914.

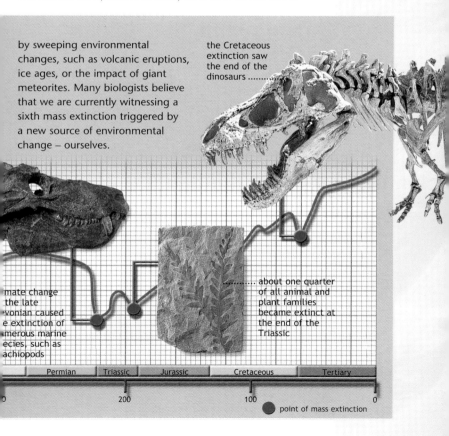

by sweeping environmental changes, such as volcanic eruptions, ice ages, or the impact of giant meteorites. Many biologists believe that we are currently witnessing a sixth mass extinction triggered by a new source of environmental change – ourselves.

the Cretaceous extinction saw the end of the dinosaurs ....

mate change the late vonian caused e extinction of merous marine ecies, such as achiopods

.... about one quarter of all animal and plant families became extinct at the end of the Triassic

| Permian | Triassic | Jurassic | Cretaceous | Tertiary |

0   200   100   0

● point of mass extinction

evolution in focus

I n a Gallup poll conducted in the US in 2001, 45 per cent of those questioned believed that God created humans in their present form at some point in the last 10,000 years. In some states, the "E" word is avoided in science classes, and in others, attempts have been made to wipe it from the curriculum altogether. There are still plenty of people who find evolution objectionable – particularly when it deals with the touchy subject of how life began. However, despite strenuous attempts to discredit it, evolution shows no signs of going away. Instead, it has become an integral part of over a dozen branches of science, from biochemistry to anthropology. Like all scientific knowledge, ideas about evolution are constantly updated by new research. Nearly 150 years after Darwin, that research is throwing new light on the way evolution works, and its path through the distant past.

> **A curious aspect of the theory of evolution is that everybody thinks he understands it.**
>
> Jacques Monod,
> Nobel Prizewinner, 1974

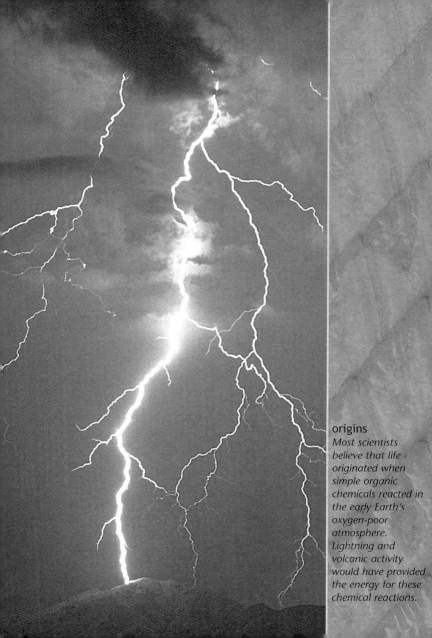

origins

Most scientists believe that life originated when simple organic chemicals reacted in the early Earth's oxygen-poor atmosphere. Lightning and volcanic activity would have provided the energy for these chemical reactions.

select company

Richard Dawkins (b.1941) is one of the world's foremost evolutionary biologists. He applied concepts originally used to describe aspects of animal behaviour – competition, cooperation, and communication – to genes, the fundamental units of inheritance. His theories have been widely popularized.

When natural selection is at work, what is it that actually gets selected? The answer – according to leading zoologist Richard Dawkins – is individual genes. In this view of evolution, living things are simply vehicles that genes use for their own ends, and all the adaptations that can be seen in nature, from stripy fur to wings, are tools for maximizing different genes' chances of being handed on. It is an idea of seductive simplicity, and one that helped to make Dawkins' book *The Selfish Gene* an international bestseller. But is it actually true?

self-sacrifice?
Ants are social insects that seem to display extreme altruism. Many individuals in a colony have sacrificed the ability to reproduce and spend all their energy sustaining a queen which is responsible for passing on the colony's genes.

group selection

Some biologists are convinced that it is, but many disagree. According to the doubters of the "selfish gene" approach, evolution cannot be reduced to a simple genetic free-for-all in which every gene is out for itself. This is because genes do not act as independent units. Instead, they are grouped together in individual organisms, which have a complex mixture of characteristics. Natural selection acts on the whole individual, rather than on its component genes. But for some biologists, selection does not stop there. According to advocates of "group selection", it also acts on whole populations, and perhaps even at higher levels too.

Until the 1960s, group selection was an accepted part of evolutionary theory. It was invoked to explain adaptations that appear to evolve "for the good of the species", rather than for the benefit of single individuals. One of these adaptations is sexual reproduction. Many single-celled organisms and plants reproduce asexually – they simply divide in two, producing two genetically identical offspring, and pass all their genes to the next generation. But with sexual reproduction, there is a genetic price to pay: on average, each partner hands on only 50 per cent of their genes. Furthermore, finding a suitable partner can be time-consuming and even dangerous – another cost that has to be factored in. So what is the adaptive reason for sex? What is its advantage? The answer is that sexual reproduction generates genetic variety, so species that use sex can adapt faster to changes in their environment. At the level of a group or a species, sex is an excellent survival mechanism.

getting close
For male snakes, sexual reproduction can be a risky business, because it often involves approaching a large and potentially dangerous mate. Courtship behaviour enables males to defuse the female's aggressive instincts, so that mating can take place.

fair shares

If natural selection really does operate at group level, it is not hard to understand why sex evolved. But there is a problem with this idea. By definition, groups are less numerous than individuals, and their "turnover" time is much slower. According to sceptics of group selection, this means that individual selection will always be a far more significant force. Like castaways fighting over limited food, individuals will always put their own selfish interests first.

*" We are survival machines –
robot vehicles blindly programmed
to preserve the selfish molecules
known as genes. "*

Richard Dawkins, 1976

In nature, this kind of selfish behaviour is never hard to find. It happens whenever one individual uses resources and denies them to another of its kind. But in some unusual circumstances, individuals really do seem to have the interests of others at heart. For example, ants work tirelessly to support their queen and raise her young, while other social animals warn their comrades of danger – even when it means running the risk of being spotted and eaten themselves. However, critics of group selection point out that this behaviour is actually selfishness in disguise. Members of these groups invariably turn out to be closely related to one another: by acting altruistically, they are simply helping to pass on the genes they share.

blood ties
*These human red
blood cells are
infected by malaria
parasites. Some
biologists argue
that the parasites
behave selflessly
because they stop
short of killing their
host, which survives
and allows the
parasites to spread,
so benefiting the
group as a whole.*

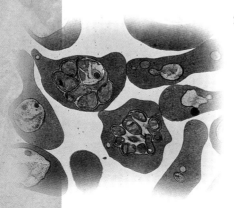

selection levels

For Richard Dawkins, this reciprocal process – known as kin selection – explains every act of apparent selflessness that is shown by living things. But after several decades in which "selfishness" has been in the ascendant, there are signs that group selection

kin selection

Closely related organisms share a high proportion of their genes, and the best way for an individual to spread its own genes may be through helping its relatives to breed. This seemingly altruistic behaviour evolves through a process called kin selection, which operates in tightly-knit groups and extended families, from elephants to colonial birds, such as the kookaburras and bee-eaters. In most of these species, young adults usually go through a lengthy "helper" stage, tending to the nests of their relatives, before eventually breeding themselves.

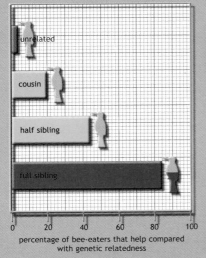

bee-eaters

Bee-eaters help to feed the young of other individuals in the colony. The degree of relatedness (shown by shading in each silhouette) dictates the percentage of individuals that help with each brood. Helpers spend more time at nests of closely related birds.

is making a recovery. Some of the theoretical debunking from the 1960s has itself been debunked, and a number of studies have shown what looks very much like adaptation "for the common good". Some researchers claim that they have found evidence of entire ecosystems working as units in natural selection – a situation that maximizes the benefits for all the species aboard.

tracking the past

Piecing together the path of evolution is a task of enormous complexity, and the further back in time you go, the harder it gets. Until the late 1960s, researchers had just two lines of evidence to help them in their work – living organisms and fossils. But following the discovery of the structure of DNA, and the deciphering of the genetic code, novel molecular techniques emerged that could help in this detective work. In many cases, these molecular studies have confirmed existing ideas, but in some instances they have led to major redrawing of evolutionary trees.

Molecular analysis works because genes evolve. Over time, mutations build up, so that copies of the same gene (and thus the proteins that the gene encodes) inherited by different species come to vary more and more. To assess how long ago two lines split apart, scientists isolate the shared genes (or proteins) and gauge the number of differences they show at the molecular level. This figure is then divided by the average mutation rate, giving the time at which the two diverged.

genetic archaeology
In 1984, scientists extracted fragments of DNA from a quagga, a relative of today's zebras, that became extinct more than a century ago. Since then, DNA has been recovered from a variety of extinct species, enabling their evolutionary history to be pieced together.

the cytochrome tree

One common method of molecular analysis involves a protein called cytochrome c. This first appeared over two billion years ago, and it now plays a key role in energy production in most living things. Like all proteins, it is made up of units called amino acids, arranged in a specific sequence. The instructions needed to assemble this sequence are held by genes, so variations in the sequence reflect genetic mutations that have built up over time. The results of cytochrome c comparisons are like a guide book to the history of life. Out of its total complement of about 100 amino acids, humans and chimps show no differences at all. On average, we differ from most other mammals in about 10 amino acids, and from fish in about 20. With flowering plants, the difference rises to about 40 places, and with fungi it hits nearly 50 – equivalent to almost one amino acid in two. The greater the differences, the more distantly species are related.

bat ancestry
Fruit-eating bats and insect-eating bats differ in many ways. In the 1980s, some experts suggested that they arose quite independently, which would explain their different features. Since then, biochemical evidence has decided the issue: bats belong to a single family.

analysing DNA

Compared to physical studies, protein analysis is an amazingly versatile way of looking into the evolutionary past. But it does have some drawbacks. One is that the average mutation rate, or "evolutionary clock", does not tick at a constant speed, particularly when different forms of life are being compared. It is a tricky problem, and one that affects chemical techniques as a whole. Protein analysis is also not very sensitive to short-term change – for example, the cytochrome c data gives no clues about when the ancestors of humans and chimps diverged. A much more discriminating method involves comparing DNA, the chemical that forms genes themselves. Strands

the moa and the kiwi

New Zealand is famous for its flightless birds. The kiwi is a living example, but much bigger species – called moas – roamed New Zealand until a few centuries ago, when human settlers brought about their demise. Ornithologists have always assumed that the kiwi is a "miniature moa", but new genetic evidence

MOA

KIWI

based on DNA collected from moa bones suggests that kiwis and moas may not be directly related at all. Instead, the kiwi shows closer links with cassowaries, birds that live in Australia. According to the DNA evidence, kiwis arrived in New Zealand much later than moas. They probably flew in, via oceanic islands, and then slowly became flightless in their new home.

haemoglobin
This image shows the structure of haemoglobin, the oxygen-carrying protein found in human red blood cells. Haemoglobin is extensively used in evolutionary research.

of DNA contain instructions for making many proteins, not just one, which greatly increases the number of differences that can build up over time.

evolution exposed

These DNA studies have been used to tease apart some of the most thorny questions about evolution's past. In particular, they have helped to clear up many of the false leads caused by convergence (see p.28) – something that often creates headaches when species are compared by physical features alone. They have been used to generate family trees for all kinds of organisms, from bacteria and fungi to songbirds and plants. This kind of research also throws light on the evolution of our own species – particularly the question about how and when modern humans arose.

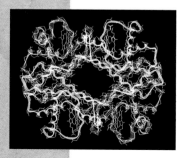

classifying living things

In the early days of natural science, the aim of classification was simply to catalogue clearly the richness of life. But since the "discovery" of evolution, classification has mushroomed into a vast biological filing system that reflects how closely different species are related. In this system, all the world's known species – currently about two million – each have their own individual place, and each one is designated by an internationally recognized scientific name. By convention, related species are grouped together into genera, and related genera into families, which are in turn members of larger categories (taxa) – order, class, phylum, and kingdom. Most biologists recognize five kingdoms, but biochemical research has also identified some even larger divisions in the natural world, known as superkingdoms. Two of these – the Archaea and

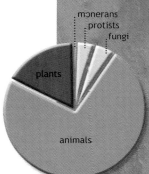

monerans
protists
fungi
plants
animals

the five kingdoms
There are far more species of animal than of any other type of living thing. However, bacteria, which belong to the kingdom Monera, easily outnumber everything else in terms of numbers of individuals.

microlife
The vast majority of living things are single-celled microorganisms. This one is a desmid – a member of a group of algae that are common in peat bogs and ponds. Its cell is divided into two matching halves.

classification and the past

Classification is a vital tool in assembling phylogenies, or evolutionary "family trees". A traditional phylogeny – like the one shown below – shows the order in which groups are thought to have diverged, and the time when these splits occurred. This information is gathered from existing species and fossils. If the evidence is clear-cut,

drawing up a phylogeny is relatively easy. But if it is ambiguous, interpretation becomes increasingly important, and different researchers may produce different results. Cladistics is a classification method that tries to minimize subjective judgments. In cladistics, species are classified strictly according to the number of "derived" features that

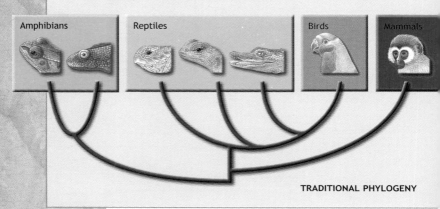

Amphibians　Reptiles　Birds　Mammals

TRADITIONAL PHYLOGENY

Eubacteria – contain fundamentally different forms of bacteria. The third, called the Eucaryota, contains all other kinds of life, including ourselves.

If biologists could see into the past, organizing this system would be a fairly routine task, albeit on a giant scale. But experts in classification – known as taxonomists – do not have this luxury. Instead, they have to deduce relationships from physical and biochemical evidence. This allows them to construct phylogenies, or family trees, which show the path evolution has followed. In this work, convergence (see p.28) is a major problem,

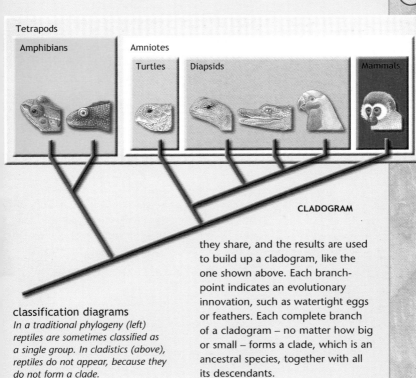

Tetrapods

Amphibians

Amniotes

Turtles

Diapsids

Mammals

CLADOGRAM

classification diagrams
In a traditional phylogeny (left) reptiles are sometimes classified as a single group. In cladistics (above), reptiles do not appear, because they do not form a clade.

they share, and the results are used to build up a cladogram, like the one shown above. Each branch-point indicates an evolutionary innovation, such as watertight eggs or feathers. Each complete branch of a cladogram – no matter how big or small – forms a clade, which is an ancestral species, together with all its descendants.

because it can make species look more closely related than they really are. To make matters more difficult, it operates at all levels within the living world. For example, no modern taxonomist would classify whales as fish, as people did several centuries ago. This is because whales have a host of features that show that they are mammals, despite their fish-like shape. But if convergence involves species that are related to begin with, relationships are harder to unravel. This explains why experts often differ about the details of classification and why classification schemes keep changing in the light of new research.

origin of life

Logic suggests that if you wind back the evolutionary clock far enough, you must reach the time when life itself first appeared. So how did this happen? One possibility is that the Earth's first living things arrived already formed from space, but most scientists are unconvinced by this idea, and believe that life on Earth originated from non-living matter. According to this view, living things developed through a long series of random chemical reactions – ones driven by terrestrial energy, such as lightning, and also by energy from the Sun.

All the essential chemical processes that characterize life take place in watery solutions, and it is most likely that water is where life first appeared. Scientists speculate that

life in the lab
In the 1950s, the American chemist Stanley Miller carried out an experiment to test the idea that life's basic ingredients could have arisen by chance. He flashed electric sparks through a mixture of gases similar to the early Earth's atmosphere, and after a week analysed the result. Miller found that a wide range of carbon-containing compounds had been formed. They included amino acids – the building blocks of proteins.

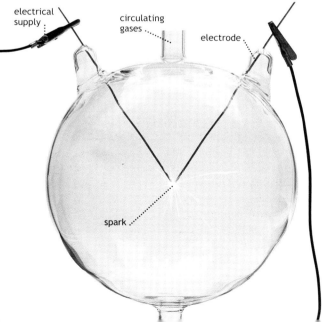

electrical supply

circulating gases

electrode

spark

the process began with simple organic substances that existed on the surface of the young Earth. Random reactions of these chemicals, over immense periods of time, produced the forerunners of living molecules. Then, at some point, a crucial event occurred. One of these chemicals started to copy itself – it became a self-

replicating system. From this moment onwards, selection and evolution began. Life was under way.

In today's living things, this self-replicating molecule is DNA, which lies at the heart of almost every cell in the body, controlling growth, development, and reproduction. But DNA almost certainly was not the first self-replicator because its copying system is highly complex, and many scientists believe that a simpler nucleic acid called RNA originally assumed this role. The very first cells probably consisted of a molecule like RNA, enclosed by some kind of protective membrane.

an improbable existence

Sceptics argue that life could not have appeared like this, because the odds against these molecular events taking place are simply too vast. But there is one observation that strongly supports the theory: all organisms alive today share the same genetic code – in other words they use the same combinations of chemical "letters" to store information in DNA. This code is one out of 10^{70} possibilities, most of which would have worked just as well. So why is there just one today? Many experts think it is because the code is a "frozen accident" – one that dates back to the dawn of life itself, about four billion years ago.

driving forces
The chemical reactions that build complex molecules need energy to take place. On the early Earth, this came from the Sun, lightning, and also from volcanic activity, which was much more intense than it is today.

playing dice
The genetic code of all living things is expressed in combinations of just four letters – C, G, A, and T (see p. 15).

human origins

Human evolution is one of the most hotly debated areas of biology. At one time, paleoanthropologists saw it as an almost linear process, with relatively few branches in the human family tree. But over recent years, new fossil finds have shown that as many as half a dozen hominid species may have been alive at any one time. Remarkably, out of the entire hominid family, only we are left.

the hominid family

Hominids came into being when our ancestors finally parted company with the apes, an evolutionary split that is now dated at between 6–8 million years ago. A host of fossil remains makes it clear that hominids originated in Africa, and then later spread to other parts of the world. The first members of the family – belonging to the genus *Ardipithecus* – had many ape-like features,

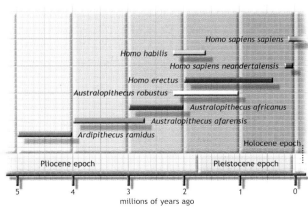

Homo sapiens sapiens
Homo habilis
Homo sapiens neandertalensis
Homo erectus
Australopithecus robustus
Australopithecus africanus
Australopithecus afarensis
Ardipithecus ramidus
Holocene epoch
Pliocene epoch
Pleistocene epoch

5   4   3   2   1   0
millions of years ago

cave art
Art is a uniquely human attribute, first appearing about 35,000 years ago. Body ornaments probably evolved as a sign of status, but paintings of animals were almost certainly associated with hunting.

but the australopithecines, which followed them, had an upright stance, together with slightly larger brains. *Homo habilis*, which appeared about 2.4 million years ago, marked the start of the line leading directly to ourselves. It was the first known toolmaker, and probably the last of our ancestors to live exclusively in Africa. By contrast, its successor, *Homo erectus*, spread right across Europe and Asia as well. Appearing about two million years ago, this species had much better toolmaking skills; and ash from fossil sites suggests that it also used fire. There is little doubt that our species, *Homo sapiens*, evolved from *Homo erectus*. The question is, how?

new features
Modern humans differ from their ancestors in having more delicate features, thinner bones, smaller jaws, and flatter faces.

competing views

There are two competing explanations for this key step in human evolution. According to the multiregional hypothesis, modern humans evolved from *Homo erectus* simultaneously in several different parts of the world. Normally, this kind of process would produce several separate species, but multiregionalists believe that early humans often mixed, preventing local species forming. The "out of Africa" hypothesis takes a quite different line. It suggests that modern humans evolved in Africa, and then migrated to other parts of the world, replacing the hominids that were already there.

mitochondrial clock

In resolving this question, evidence from living humans is just as important as evidence from fossil remains. This modern data comes from small loops of DNA found in mitochondria – the self-contained power plants found in living cells. Unlike the DNA in cell nuclei, mitochondrial DNA (mtDNA) is inherited intact from the mother alone. As a result, the only changes are random mutations, which build up over time. Since the late 1980s, attempts have been made to read this mtDNA clock, to put a date on the most recent shared ancestor of all the people living on Earth.

there and back

So far, mtDNA data indicate that all modern humans originated in Africa between 140,000 and 300,000 years ago. This is strong backing for the "out of Africa" hypothesis, because by then, *Homo erectus* had already spread to other continents. Some researchers have suggested that modern humans emigrated from Africa in a single wave, perhaps as little as 50,000 years ago, but according to some recent genetic studies, there might

Lucy
Discovered in Ethiopia in 1974, these are the 3 million-year-old fossilized remains of Lucy, a female Australopithecus afarensis. *Her skeleton shows that she weighed less than an adult chimpanzee.*

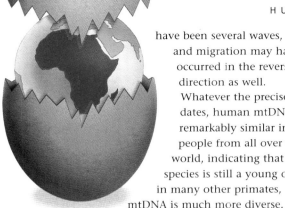

have been several waves, and migration may have occurred in the reverse direction as well. Whatever the precise dates, human mtDNA is remarkably similar in people from all over the world, indicating that our species is still a young one – in many other primates, mtDNA is much more diverse.

out of Africa
Early Homo sapiens *originated in Africa and quickly occupied the European and Asian continents. They reached the Americas by crossing the Bering Strait at a time when sea levels were about 100m (330ft) lower than today.*

then there was one

If the "out of Africa" hypothesis is true, then what happened to the hominids that were already in Europe and Asia when modern humans arrived? At the time, these inhabitants were the Neanderthals – heavily built hominids that had larger brains than modern humans. Some paleoanthropologists believe that Neanderthals and modern humans interbred, but the majority suspect that they suffered a bleaker fate. Despite their superior brain size, they were technologically outclassed by the newcomers. For example, it seems that Neanderthals never invented weapons that were designed to be thrown. After thousands of years of being marginalized, they eventually became extinct.

Mary Leakey (1913–1996) was one of the 20th century's greatest paleo-anthropologists. Her early finds, such as *Australopithecus boisei* – one of the so-called robust australopithecines – established her reputation. In 1978, she made a sensational discovery: two sets of australopithecine footprints (see p.7). With her husband Louis and son Richard, she was part of one of the world's most successful scientific dynasties.

evolution and the future

> **" Man with all his noble qualities ... still bears in his bodily frame the indelible stamp of his lowly origin. "**
>
> Charles Darwin, 1871

After nearly four billion years, evolution has produced something truly unique: a species that can understand the evolutionary process, and modify its future course. This interference started when humans first domesticated plants and animals, and it looks set to accelerate enormously as we begin to alter the genomes of living things – including ourselves.

Humans are still subject to natural selection, but its command over us is fading fast. We escape some of its effects through modern medicine, and also through technological aids that keep natural hazards at bay. But this freedom from selection comes at a price, because it means that potentially damaging genes are no longer weeded out. As a result, humanity's future gene pool will contain a growing proportion of deleterious genes – ones

cultural evolution
The human ability to pass on learned information is unique in the living world. As new forms of communication are developed, this store of information has been evolving at an increasingly rapid rate, leaving our physical evolution trailing far behind.

that, ironically, medicine has helped to maintain. In the future, this problem will be targeted by a new kind of medicine – one that deals with genes themselves. As human lifespans increase, genetic medicine will tackle all kinds of disorders, including ones triggered by genes that become active late in life. Because these genes "switch on" after people have finished reproducing, natural selection cannot curb them at all. Instead, genetic medicine will take its place.

Physically, the human species has changed relatively little in the last 10,000 years. But during the same time, the combined store of human knowledge – our culture – has evolved enormously, and in a non-Darwinian way. Already, this avalanche of knowledge is enabling us to tackle some of our physical shortcomings, and to tailor other living things to meet our own ends. In the long term, it will go much further, creating extraordinary possibilities for human beings.

changing fortunes

In the non-human world, adaptability will become increasingly important for biological success. As humans continue to change the planet's climate and habitats, species that can cope with change will thrive, but ones with highly specialized lifestyles will not. This is good news for many of today's urban animals, and also for plants that live on disturbed ground. But for large numbers of species, from mountain plants to giant pandas, the future is not so bright.

hidden traces
The human mind still bears the imprint of our past. Evolutionary psychologists believe that many of our behavioural responses – from sex to shopping – can be explained in terms of adaptations that evolved long ago.

endangered species
Faced with shrinking natural habitats, hunting, and other threats, many of today's large mammals risk becoming extinct in the wild. Ironically, their long-term survival will ultimately depend on human help.

glossary

adaptation
An inherited change that improves an organism's chances of survival, and of leaving viable offspring.

adaptive radiation
The diversification of an original species into a group of descendants, each adapted for particular ways of life.

adaptive value
The value of an adaptation, in terms of its ability to help an organism to leave viable young.

amino acid
A nitrogen-containing compound that living things use as a building block to make **proteins**.

anagenesis
A form of **evolution** that changes one species into another, without creating any additional **species** at the same time.

analogous structure
A structure or characteristic shared by two **species** that has evolved in two different ways, from two different starting points. *See also* **convergence**.

artificial selection
Deliberate selection of organisms by human beings, carried out to emphasize useful or desirable characteristics.

asexual reproduction
A form of reproduction that involves a single parent, without sexual fertilization taking place. In most cases, it produces offspring that are genetically identical to their parent.

chloroplast
A structure inside plant cells that captures the energy in sunlight, and turns it into a chemical form.

chromosome
A package of DNA and protein. In **eukaryotes**, chromosomes are normally present in pairs, except in cells that are used for reproduction.

clade
An ancestral species, together with all its descendants, whether existing or extinct.

cladistics
A method of classification based on shared derived features, which are ones that have appeared relatively recently in an organism's evolutionary past. In cladistics, closely related species form **clades**.

coevolution
Evolutionary change in one **species** that is linked to changes in another. Coevolution often produces increasingly specialized ways of life.

commensalism
A relationship between two species in which one benefits, but the other neither benefits nor suffers.

convergence
A process by which unrelated species develop similar **adaptations** in response to similar conditions.

Darwinism
The view put forward by Charles Darwin, that species evolve as a result of **natural selection** acting on inherited variations. The modern interpretation, known as neo-Darwinism, holds that **evolution** occurs because of changes in gene frequencies, caused mainly by natural selection.

directional selection
A form of selection that creates a trend in one direction, by favouring individuals at one extreme.

disruptive selection
A form of selection that favours individuals with different forms of a characteristic, eliminating ones that are intermediate between the two.

DNA
Deoxyribonucleic acid. In all organisms, apart from some viruses, DNA is the ultimate source of genetic information.

eukaryote

An organism that has complex cells, containing nuclei and a range of functional units, or organelles, that carry out different tasks.

evolution

A gradual change in the genetic makeup of a group of living things. Evolution occurs over successive generations, rather than in the lifetime of individuals, and is a continuous process that cannot be reversed.

extinction

The permanent disappearance of a species, or of a group of species.

fitness

The relative success of an organism, measured by its ability to leave viable young.

gene

A unit of inheritance. A gene is a specific sequence of DNA that contains the information needed to make a single protein.

group selection

A form of selection that operates at a group level, rather than on individuals. In group selection, characteristics may evolve that benefit the group, but not necessarily the individuals that make it up.

hominid

A member of the primate family that includes humans, together with all our immediate ancestors, but excluding apes.

homology

A shared "design" or body plan seen in related species that is the result of their common ancestry.

isolating mechanism

A biological barrier that tends to prevent successful interbreeding by members of two groups, so allowing new species to evolve.

kin selection

A form of selection that acts on groups containing closely related individuals. Kin selection can favour behaviour that is not advantageous to individuals living on their own.

Lamarckism

An evolutionary theory proposed by Jean-Baptiste de Lamarck, which holds that characteristics developed or lost during life can be passed on.

macroevolution

Major evolutionary change, involving groups above the level of species. This kind of evolution often involves visible changes in form.

mass extinction

The collective extinction of a large number of species within a relatively short space of time.

microevolution

Minor evolutionary change that modifies existing species, without creating new ones. At a molecular level, microevolution is continuous, but often has no visible effects.

mitochondrion

A microscopic structure found in cells, which liberates energy by oxidizing a chemical fuel. In animals, this fuel is usually glucose derived from food.

mutation

Any change that affects the genetic information held by DNA. Mutations are accidental events, and they are responsible for generating the genetic variations that allow living things to evolve.

mutualism

A partnership between two species in which both benefit. *See also* symbiosis.

natural selection

A passive process of selection that screens variations in living things. Natural selection favours variations that maximize an organism's ability to leave viable young, and penalizes ones that reduce reproductive success. Unlike artificial selection, natural selection works through a host of environmental factors. Some stem from an organism's physical habitat, and some from the other organisms around it.

nucleic acid

An organic molecule that is used to store genetic information, or to put it into effect. In most living things, apart from some viruses, the storage molecule is DNA.

organic
In its chemical sense, an organic substance is one that contains carbon. In biology, it refers to any constituent or product of living organisms.

parasitism
A long-term association between two species in which one, the parasite, obtains its nutrients from another, known as its host.

phylogeny
The evolutionary history of a group of living things. Phylogenies are often expressed graphically, as evolutionary trees.

polyploidy
A permanent change in the number of chromosome sets found in living cells. Once the change has occurred, it may be handed on when an organism breeds.

preadaptation
An **adaptation** that evolved in response to one set of conditions which later turns out to be useful in another.

predation
A way of life followed by many animals that involves catching and eating prey.

protein
A molecule consisting of a sequence of **amino acids**, arranged in a particular order. Proteins are the workhorses of living cells: they carry messages, speed up chemical reactions, and form many of the structural components of living things.

punctuated equilibrium
A view of **evolution** that holds that periods of rapid **macroevolutionary** change are interspersed with longer periods of stasis, or equilibrium, in which no **macroevolution** occurs.

reproductive isolation
The physical or biological isolation of a group of organisms, which prevents them from breeding with others of their own kind.

selection
All the factors that alter the balance of **genes** as one generation of organisms succeeds another.

sexual reproduction
Reproduction that involves male and female sex cells, which join together at fertilization. Unlike **asexual reproduction**, sexual reproduction creates offspring that are genetically different from their parents.

sexual selection
A form of selection that occurs when individuals of one sex compete among each other for mates. Sexual selection can encourage features that have no useful function, apart from attracting the opposite sex.

speciation
The **evolution** of new **species**. Speciation can create a new species from an existing one, or it may transform one species into several descendants.

species
A group of genetically similar organisms that breed exclusively or largely with their own kind. The species is the fundamental unit of **taxonomy**, and it is the only taxonomic group that has a concrete existence in the natural world.

stabilizing selection
A form of selection that tends to favour individuals with average characteristics, penalizing those that lie towards the extremes.

symbiosis
Any kind of close partnership between two different **species**. Frequently, the term is used incorrectly to mean a mutually beneficial partnership.

sympatric speciation
The evolution of one or more new species from populations of a single original species living in the same geographical region.

taxonomy
The science of identifying living things, and of classifying them according to their evolutionary past.

trace fossil
Indirect signs of living things which have been preserved as fossils.

vestigial organ
An organ that has become redundant as a result of evolutionary change. Vestigial organs tend to be much smaller than ones that are fully functional.

index

Further reading

Almost Like a Whale: the Origin of Species Updated
Steve Jones, Black Swan, 2001.

The Blind Watchmaker
Richard Dawkins, Penguin Books, 1990.

The Origins of Life
John Maynard Smith and Eors Szathmary, Oxford Paperbacks, 2000.

The Diversity of Life
Edward O. Wilson, Penguin Books, 2001.

The Third Chimpanzee
Jared Diamond, HarperCollins, 1992.

Dawkins Vs. Gould Kim Sterelny, Icon Books, 2001.

Evolution Mark Ridley (ed), Oxford Paperbacks, 1997.

What Evolution Is Ernst Mayr, Weidenfeld, 2002.

Useful web addresses

Human origins at the Smithsonian Institution
http://www.mnh.si.edu/anthro/humanorigins/

The Charles Darwin Foundation
http://www.galapagos.org/about.html

The Tree of Life Web Project
http://tolweb.org/tree/phylogeny.html

Author's acknowledgments
I would like to thank Hazel Richardson of Dorling Kindersley for her input in the preparation of this book, Marek Walisiewicz, of cobalt id, for his informed and adroit editing, and all those scientists and teachers who are currently defending the public's right to learn about Darwin's "great idea".

Jacket design
Nathalie Godwin

Illustration
Halli Verrinder

Index
Indexing Specialists, Hove

Proofreading
Dr. Kim Bryan

Picture credits
Bruce Coleman Collection: Bridgeman Art Library 14; Johnny Johnson 24(br); Kim Taylor 26; HPH Photography 51; Pacific Stock 59. FLPA: Lanting 5; F. Bavendam 17; Natura Stock 19(br); M. Iwago 30(bc); Steve Maslowski 35. Oxford Scientific Films: Richard Herrmann 10; Martyn Colbeck 21(bl). RSPB Images: 20(br), 32. Science Photo Library: John Reader 7(tr); Volker Steger 15; Dr Kari Lounatmaa 23; Noah Poritz 31(br); Dr Kari Lounatmaa 33; CNRI 39; David Parker 43; Keith Kent 47; Dr Gopal Murti 50; J.C. Revy 54; Keith Kent 61(t); John Reader 62, 63; Mehau Kulyk 65; Marc Atkins 48.

Every effort has been made to trace the copyright holders.
The publisher apologizes for any unintentional omissions and would be pleased, in such cases, to place an acknowledgment in future editions of this book.

All other images © Dorling Kindersley.
For further information see: **www.dkimages.com**